Anonymous

A Short Attempt to Recommend the Study of Botanical Analogy

Anonymous

A Short Attempt to Recommend the Study of Botanical Analogy

ISBN/EAN: 9783337175559

Printed in Europe, USA, Canada, Australia, Japan

Cover: Foto ©berggeist007 / pixelio.de

More available books at **www.hansebooks.com**

A

SHORT ATTEMPT

To recommend the

STUDY of BOTANICAL ANALOGY.

A
SHORT ATTEMPT

To recommend the

STUDY

OF

BOTANICAL ANALOGY,

In inveſtigating the

PROPERTIES of MEDICINES

FROM THE

VEGETABLE KINGDOM.

Si quid dixero forte jocoſius, hoc mihi juris
Cum venia dabis.   Hor.

LONDON,

Printed for G. Robinson, Paternoſter-Row;
and C. Elliot, Edinburgh.

MDCCLXXXIV.

TO

Dr. JOHN HOPE,

PROFESSOR of BOTANY and MEDICINE, in the Univerſity of EDINBURGH.

SIR,

IT would be a ſufficient excuſe for my preſuming to aſk your protection ont his occaſion, that, from you, both my knowledge of botany, and zeal to promote it, were originally

nally derived. If the firſt efforts are weak and unſucceſſful; if the firſt fruits are crude; time and attention may beſtow greater ſtrength, and a more complete maturity. You have already felt the pleaſure of foſtering and protecting the tender plant, and may recline under extenſive ſhades, where you once found a comparative deſart. May this be again your lot, for the author can truly ſay,

" 'Tis but to try his ſtrength, that now he ſports,
With Chineſe gardens, and with Chineſe courts!"

This

This addrefs is however
more properly yours for another reafon. It is well known
that a natural method, the
" primum & ultimum in botanicis defideratum," has been
particularly the object of your
attention; and that, from you,
we may expect fomething more
complete than " fragments."
If therefore the knowledge of
the Materia Medica is improved
from an attention to Botanical Analogy, the improvement
will be more complete, as a
natural method is brought to
greater

greater perfection.   From you then we expect it ; and from you, we shall receive it with gratitude.

I am, with great esteem,

Your most obliged, and

most humble servant,

Paternoster-Row,
Sept. 14 1783.

The AUTHOR.

A

SHORT ATTEMPT

To recommend the

STUDY of BOTANICAL ANALOGY.

THE moft refpectable phyficians have frequently complained, that our medicines were too numerous; and have joined in wifhing, that thofe which were lefs efficacious fhould be feparated from others, whofe effects were more confiderable and certain. The different colleges, in their new Pharmacopoeias, have diminifhed the number

B   of

of officinal remedies, and rendered their compofitions more fimple as well as more elegant;—but, like the heads of the fabulous Hydra, the endeavours to retrench occafion an increafe. New medicines are frequently ufed, fupported by the moft extravagant commendations, and offered to the public as poffeffed of powers little fhort of infallible. If thefe praifes were well founded, both fcience and mankind might boaft of the improvement, and the materia medica would foon confift of a fhort catalogue of efficacious fimples. It were an infult however on the reader, to recall to his remembrance the number of new remedies which were known only to be

again

again neglected; and it might occasion the unpleasing recollection of his disappointments, when he had built his hopes on the very uncertain foundation of specious promises.

In the hours of leisure and retirement, the author of this tract was induced to review the different remedies which had received their portion of applause, and were consigned to oblivion. It was not the work of an idle curiosity; for it was probable that, in this neglected lumber, something valuable might still be discovered. Philosophers are often fickle in their attachments, but the remedy which had been for ages valued, was not probably destitute of merit. From the examination of

B 2

some

some neglected authors, which the courtly elegance of modern times turns from with contempt, he was induced to confider the different methods of inveftigating the powers of medicines, and to wade through the mafs, which folly, fancy, and fuperftition, had collected, in expectation of fome particles of gold. It is not the intention of the prefent work to purfue this laborious tract ; but in the more luminous period of modern times, he was particularly attracted by the botanical analogy, and the remarkable agreement between the natural characters of plants, and their effects on the human body. Though there muft necefsarily be many exceptions, yet

it

it appeared the moft ready method
of difcovering the probable effects
of a new remedy, and fometimes of
checking the ardor of innovation,
the fafcinating recommendations
which it had dictated, and a fondnefs
for novelty had enforced. It is therefore the defign of the prefent fhort
attempt, to vindicate this method of
inveftigation, which does not require the fmoaky labour of the fur-
nace, — the patient and painful,
though fallacious attention to phials
and mixtures, or the danger of a per-
fonal experiment. It will be a fufficient recommendation, if we find, in
many of the claffes of vegetables, a
very general relation of their feveral
powers, and, in the boafted novel-

ties

ties of modern times, remedies fimilar to thofe which have been alternately admired and forgotten—cultivated and neglected.

In the prefent attempt, indeed, there are few pretenfions to novelty. The famous Englifh botanift, Mr. PETIVER fuggefted this method in the Philofophical Tranfactions; the laborious and accurate HOFFMAN purfued it; and the difciples of the LINNÆAN fchool have endeavoured to perfect it. Their attempts are refpectable, and fhall not be neglected in the prefent view of the fubject: but much will ftill remain; and thofe, who know what they have performed, will be the beft and moft candid judges of what has been added.

At

At the prefent era of Botanical knowledge, it might be prefumed, that the term GENUS is fufficiently underftood ; yet, though obvious, it has been mifapprehended ; and though fimple, mifreprefented. Linnæus, to whom the world is indebted for the accuracy with which the numerous fubjects of the vegetable kingdom are diftinguifhed, has felt the vengeance which difappointed ambition can inflict. His works have been ftigmatized as a grammar and a dictionary, when, in fact, he aimed at no more ; and he has been accufed of ftopping the ftudent at the threfhold of fcience, both by the ufelefs obfcurity of his language, and confining the views

of

of the naturalift to diftinction only..
Thefe accufations have been fatif-
factorily anfwered in other places ;
and they would not now have been
introduced,. were it not to ftate, in
oppofition to them, one of the nu -
merous advantages which his labours have beftowed. It will be
obvious that SPECIES only exift in
nature ; the various hues of the flow-
er, the fize and ramifications of the
branches, are frequently changed by
the foil and climate :—they are the
fports of chance, for the vegetable,
in its proper fituation, returns to its
former appearance. The firft and
moft natural arrangement of the
fpecies forms what botanifts have
ftyled a GENUS. It ought indeed to
be

be ftrictly natural; but, as the fpecies are fo numerous, an inconfiderable licence has been allowed in
this refpect, in order to abridge the
number of genera. Linnæus, who
had examined plants with the moft
accurate and unwearied attention,
found fome reafon to make them ftill
more comprehenfive, and to feparate
thofe of other authors, that he might
form his genera in a more natural
manner. This is the proper criterion of the merit of a naturalift;
but, unfortunately, his numerous
antagonifts have been unwilling or
unable to arraign his conduct in
this refpect. The patient and cautious philofopher fometimes finds
reafon to queftion the propriety of
his

his conduct, but the same knowledge, which points out the apparent error, suggests the apology; viz. the amazing variety of nature, and the almost insuperable difficulty of confining her within the limits of a system. It has thus happened that the admirers of the Swedish naturalist have been distinguished for the extent of their acquisitions; and in the later period, when the terrors of innovation have subsided, his enemies have been only the vain—the ambitious—and the superficial.

It was necessary to state this imperfection, even in the first and apparently the most easy attempt to arrange the subjects of the vegetable

able kingdom, becaufe it might, with fome plaufibility, have been urged againft any argument which would derive the virtues of plants from their botanical analogy. But the objection would have been only plaufible. Though the genera be in fome degree artificial, it very feldom happens that the virtues of the fpecies materially differ, except in degree. All the fpecies of the Rhubarb are both purgative and aftringent. The Cincona Caribbæa is a tonic, as well as the C. officinalis, and probably equally certain. All the fpecies of the Allium poffefs the peculiar properties of Garlic. It would be endlefs to purfue this matter in all its varieties ; yet

it

it is neceffary to add one fact, which will clearly evince the propriety of attending to the genus; and it will equally fhow that foil and climate make a very flight alteration in the medical properties of the vegetable. The Seneka or rattle-fnake root was much valued by the original inhabitants of Virginia, for its good effects in curing the bites of the fnake, from whence it received its name, and as a very efficacious remedy in pleurify, peripneumony, and other active inflammations. Mr. Tennent, with a very laudable induftry, difcovered the plant, and found it to be a fpecies of the Polygala. The European fpecies of the fame genus was therefore tried, and its efficacy was

was found to be very little inferior.
The Seneka has indeed loft its credit;
but the reafon is obvious : as it pow-
erfully excited vomiting, and its
confequent evacuation by the fkin,
it was very well adapted to thofe difeafes for which it was employed.
The ardour of a difcoverer overlook-
ed this very probable caufe of its
efficacy, and attributed it to a fpe-
cific quality in the ROOT ITSELF.—
Philofophy corrected the eagernefs
which had occafioned and fupported this opinion,—till reflection fuggefted that we need not ravage the
American continent, for an active
and ufeful emetic.

The genus of CONVOLVULUS affords
us a ftriking example of the medi-

cal

cal powers pervading a natural col-
lection of fimilar fpecies. From this
genus, we have the SCAMMONY—
the TURPETH—the MECHOACANNA,
the SOLDANELLA; and lately we have
found that it affords us alfo the JA-
LAP;—befides that, in its different
countries, it is the moft frequent
domeftic remedy of the native inhabitants. Another very comprehenfive genus, which Linnæus has efta-
blifhed, is the EUPHORBIA. It contains the fpecies of the original Euphorbium,—thofe of Tournefort's
TITHYMALUS, and the ESULÆ of Ri-
vinus. They are various in their habits, and external appearance; but
they are fimilar in their properties,
for they are all lactefcent, and
highly

highly ftimulant. They were formerly employed as purgatives in
dropfy; but the violence of their action has deterred modern phyficians
from their ufe. Profeffor Guilan-
dinus funk under their operation;
and, though Lifter attempted to re-
vive them, they were foon entirely
neglected. We have chiefly mentioned this genus, becaufe we know
no one that is apparently more irre-
gular and unnatural. It cannot fail
of occurring to an attentive reader
of the more ancient phyficians, that
all the old remedies for dropfies,
were of this violent kind of draftic
purgatives. The more gentle and
timid modern is terrified by their
effects, and probably often fails of
relicv-

relieving the patient, becaufe he is more afraid of the remedy than the difeafe. It is illiberal to fuppofe that the ancients were cool deliberate murderers, and purfued a deftructive courfe, without a proportional number of cures to confole them for their many loffes. When we fee, in the courfe of fucceffive ages, medicines of a fimilar nature employed, and varying only in their fource, as vegetables or minerals, we ought to allow that purgatives are frequently, and perhaps generally, beneficial. It fhould therefore be our bufinefs to afcertain, with fome precifion, the circumftances which may regulate our conduct, and to determine in what cafes, purgatives,

tives, and where the more gentle diuretics should be employed. After much attention to this point, dropsies seem to have been seldom effectually relieved without copious evacuations by stool ;—but it is also necessary to add, that this should be procured by the mildest means ; though when these are ineffectual, even the violent drastic of Dovar is sometimes of service. The attentive practitioner will always find that large and copious watery stools relieve the patient ; and, though their number may at first terrify him, yet the benefit which he daily receives will add to his confidence, both in the remedy and his physician —But we must return to our subject.

C   The

The true ACACIA of the Greeks,
the *δακρυὸν κυανωπον ἀκανθής* of Andro-
machus, has been long neglected,
and would almoſt have been forgotten, if the name had not been pre-
ſerved, by our retaining under the
ſame title, the inſpiſſated juice of
the unripe ſloes. It was confeſſed-
ly a very powerful aſtringent, frequently uſed in Egypt, both as a
medicine and an ingredient in various
œconomical preparations; but, by
the neglect of our merchants, or
the prevalence of faſhion, it has been
long ſince unknown, and we have
preferred the TERRA JAPONICA,
which has, at leaſt, the advantage
of coming to us by a longer voyage,
and

and perhaps at a greater expence.
By the care of Mr. Kerr, who has
long refided in the factory at Patna,
we have, at laft, received a defcription of the plant, from which the
juice, in its infpiffated ftate impro-
perly called an earth, is prepared ;
and we find that we have probably
recovered a very fimilar remedy to
the ancient acacia. The one is an extract from the MIMOSA NILOTICA,—
the other from another fpecies of the
fame genus. The names might have
fuggefted as much to a dexterous e-
tymologift; for *acacia* or *akatia*, without the Arabic prefix, *a*, is not very
unlike *kaath*, *cate*, and *caetchu*.—But
fortunately, amidft the various caufes
of confufion in medical enquiries,

etymo-

etymology has not found a place; phyficians are generally contented with obferving a fimilarity of facts; and when they are fo happy, are often negligent about names. It may be proper to add, that the very ufeful exudation, the GUM ARAB. proceeds from the fpecies of *mimofa*, from which the *acacia* was formerly prepared. It may therefore become neceffary to enquire, whether the gum may not be procured from other fources, fince we have now a *fhare* only of the trade, by the courtefy, and at the will of our former enemies.

It is unneceffary to inform the learned reader, that the GUM ELEMI, once highly valuable, is forgotten,

while the BALSAM OF MECCA,—
with its fruit and wood, are ftill care-
fully preferved in the Eaft, and only
refigned, through defpair of obtain-
ing them in Europe. We have been
lately informed, that both thefe
fubftances are the production of fimilar plants, included under the fame
genus. But the AMYRIS OPOBALSA-
MUM is rarely found even in Afia,
and the ELEMIFERA, though it has
been feen in Carolina and the Weft
Indian iflands, is alfo uncommon.
There is a fpecies in Jamaica which
is lefs fo ; and it may be of fervice
to try its effects. We often wander
to a great diftance in fearch of remedies, which nature profufely of-
fers at our own doors. The fpecies
C 3   which

which I would recommend to the attention of travellers, is mentioned by
many authors.—It is defcribed by
Sloane as " Lauro affinis, tereben-
" thi folio alato, ligno odorato can-
" dido, flore albo." This fpecies
certainly deferves a trial, by thofe
who are ftill willing to truft a remedy which has been fo much celebrated. Its virtues however, when
feparated from the exaggerations of
fuperftition, are probably few; it
feems to be a warm cordial and a
diuretic; but, as it has been fuperfeded by the lefs fragrant turpentines in the laft effect, the former may be eafily obtained by more
agreeable medicines.

There are ftill many old remedies

·dies which have been superseded by their kindred exotics. The ARISTO-LOCHIA LONGA and ROTUNDA are neglected, while we have lately pur-chased the SERPENTARIA from a very similar plant of the same genus, at a vast expence. A candid practitioner would be at a loss to point out the difference of their qualities, except in degree; and when he has actually tried them, perhaps would very willingly consign both the one and the other to oblivion. If we want a warm stimulant, we may readily find it in a less suspicious class.

It might be easy to extend this catalogue of foreign species, which we eagerly adopt, while our own climate furnishes others belonging

to

to the fame genus. The few that have been given, were not felected for their peculiarity, but as they might afford fome entertainment to a fpeculative mind, who furveys the equal inftability of literary fame, and of human greatnefs and power. They are fufficient to point out what is one of the chief objects of this flight attempt; that botanical labours do not end in idle curiofity alone; and that even the dictionaries of the fcience may fuggeft fubjects of ufeful fpeculation. The reveries of our countryman fir John Hill, are probably yet remembered, and may involve the prefent attempt in the fame ridicule. But our object is very different. It muft be remembered,

membered, that, at prefent, we by
no means wifh to recommend our
own fimples alone : our fuperflui-
ties enrich us, by employing our
manufacturers to procure the proper
returns; but, if remedies are procured with difficulty, and at great
expence; if, as frequently happens in fuch circumftances, their
condition is bad, and they are adulterated by cheaper and lefs ufeful
materials, we ought certainly to
turn our eyes to our own paftures,
or thofe colonies which depend on
us, and enquire how far we may
be more certainly and conveniently
affifted by them.

It is a more difficult tafk to point
out the congruity in the natural orders ;

ders; they are neceffarily more arbitrary, and the fimilarity is confequently lefs ftriking. If it be found
however that the agreement is fo
great as to enable the moft expert
Botanifts to arrange the feveral genera in the order of their powers,
they may affift the ftudent in his
progrefs, and teach us to judge of
the qualities of a medicine, when
firft offered to our notice. A complete natural method is the ulti-
mate purfuit of the Botanift, but
it is the univerfal medicine, or
fquaring the circle, a propofition
phyfically impoffible. This will be
obvious from a flight reflection : our
limits of claffes and orders are the
confequences of our imperfect knowledge ;

ledge ; nature proceeds by degrees, and from the cedar of Lebanon to the hyffop which groweth on the wall, there is no part at which the line of divifion can be drawn with propriety. In the infancy of our knowlege of nature, many families of plants were readily diftinguifhed, and they remain at prefent ufeful monuments of our prefumption and our imperfections ; for future difcoveries have filled up thofe chafms which occafioned the divifion ; and we muft either reject the plants from our fyftem, or thofe limits which we fondly imagined were dictated by nature. It would be no very difficult tafk to fhew that an arbitrary arrangement is better fuit-
ed

ed for the chief purpose of a system, viz. distinction; while the natural classes might be preserved, with all their imperfections, as some guide in the discovery of the nature and properties of vegetables. Linnæus, who has derived his greatest praise from an artificial method, yet eagerly pursued a natural one, which he calls " primum & ultimum in Botanicis desideratum;" but from its necessary imperfections has modestly styled his attempts, Fragments. It is from these fragments therefore, that our future observations will be deduced; and though there is much uncertainty, and some contradiction in several parts, yet there is sufficient information to be derived

from

from them to encourage the purfuit. It will at leaft ferve to increafe
the knowlege of nature, and to
extend our enquiries concerning the
properties of many vegetables,
which have been hitherto mifunder-
ftood or neglected.

It may be objected, that we already poffefs an extenfive catalogue
of medicines, whofe properties have
been afcertained : that thefe have
been arranged in various elegant
fyftems, which have frequently attracted the admiration of
the ftudent, and the contempt
of the practitioner. Difeafes, in
fact, are not cured by fyftems; and
thefe fplendid monuments of miftaken talents, and mifapplied in-
dustry,

duſtry, frequently miſlead us. E-
very medicine undoubtedly poſſeſſes ſome diſtinguiſhed powers, which
will ſupport the author's opinion
in its arrangement, and give the
appearance of accuracy as well as
of elegance. But there are various
circumſtances, which will influence
the choice of the medicine, from its
inferior properties : theſe are, in ſuch
ſyſtems, little attended to; and it
alſo poſſeſſes powers, for which in
emergencies it may be uſed, when
we have little choice of the methods
to be employed. Thoſe who have
found ALUM, in their ſyſtems, a-
mong the aſtringents, will not readily think of it as a purgative; and
where the POLYGALA is reckoned
only

only an expectorant, it will not be
readily ufed, even in the moft ur-
gent fituations, as an emetic. But
I am by no means willing wholly to
difcard thefe arrangements—they are
probably ufeful both to the teacher
and the ftudent; yet they may be
properly regulated, by inducing the
learner to look beyond them, and by
fhowing him that they are the means
only, rather than the end. They
give a general and comprehenfive
view, which is afterwards to be va-
ried and enlarged. They are the de-
finitions and axioms of the geome-
ter; the foundation indeed of every
aftronomical calculation, but never
thought of fince the days in which
the ftudy of Euclid commenced.
But

But whatever may be the merits of these systems, they will not preclude our present attempt. These are the result of observations already made ; this is a guide only to a cautious trial : the former is the consequence of established facts ;— the latter leads us to establish them. The imperfection of this method, for it is confessedly imperfect, has led many naturalists to suggest the trial of tastes and smell.—These therefore have been diversified beyond the powers of language to ascertain their difference ; but in their event are uncertain, and sometimes dangerous. The taste and smell, may be, with propriety, combined with the botanical analogy, but

neither

neither can entirely superfede the other.

Perhaps it may not be wholly uselefs, to give a short account of the later fystems of this kind. They posfefs a degree of clearnefs and accuracy, which will scarcely be expected in a subject so doubtful and uncertain; while, at the same time, they will, in some degree, affift the principal object of this attempt. It were eafy to fill this tract with the reveries of authors, and to copy pages of founding trifles; but, as Sterne wished to speak a few words point blank to the heart, so we shall principally aim at the underftanding. Both may fail, but the attempt is commendable.

D The

The divifion of taftes, by which we difcover a difcriminated permanent property in the body poffeffing them, is the following. The *watery* and *dry*—the *vifcid* and *falt*—the *fweet* and the *acrid*—the *acid* and *bitter*—the *fatty* and the *flyptic.* The watery tafte is well-known : the chief examples are the oleraceous plants. The *dry* is perceived in the ivy—the capillary herbs, and particularly in the feeds of the lithofpermum, which Grew informs us effervefce with acids. But from the firft divifion we gain little information. All vegetable food is gently laxative, it is alfo flightly nutricious and cooling; fo that vegetables, which poffefs few other diftin-

guifhing

guifhing properties, are sure to a-
gree in thefe. There are many,
however, which fill the herbals of
the curious, which poffefs no fuperior merit—their virtues are detailed
with care, but they are fo flight,
that we may fafely neglect them.
Their oppofites, the dry fubftances,
have ftill lefs power; indeed, except
by abforbing acrid humours on the
furface, and, in a fingle inftance,
if Grew's folitary experiment be
true, by deftroying acidity in the
ftomach, we receive little benefit
from their efficacy.

The vifcid and falt are not fo juftly oppofed. The former is exem-
plified by the gums and the mucilaginous vegetables; but their only

D 2   proper-

property is that of a demulcent.—
The common falt externally is a fimple ftimulus, internally laxative.—
It requires many other circumftances to concur with its action, before it can with juftice be accufed
of producing the fea-fcurvy and
malignant ulcers; yet there is no o-
ther example of this tafte.

The fweet tafte is fufficiently
known, but it adds little to our information; for between the praifes
and the cenfures of fugar, reafon
can with difficulty decide. It is
therefore probable, that it deferves
neither, in a confiderable degree.
In a moderate quantity it is nutritive, and, when flowly diffolved in
the mouth, demulcent; but we can
add

add little more; for all the fweet vegetables feem to poffefs no very different qualities. The laxative power of brown fugar, manna, and cafia, probably depend on the mixture of fome oily or refinous matter of the vegetable to which they belong; for it is diminifhed in the fame proportion as the purity of the fugar is increafed. Its moft proper oppofite is the acrid—the onions, fpices, muftards, and warm ftimulants of our own climate, are the examples of this tafte. They are all ftimulants, and fome of them highly ufeful in this view; fo that this quality is generally the moft exact teft of a ftimulating power. But they are alfo diftinguifhed by other

D 3   qualities,

qualities, for they evacuate the
fluids from different glands. To
properly examine this subject,
might too far extend the present
tract, but we should probably find,
that no taste of this kind will just-
ly indicate any evacuating power,
except that of onions. The whole
tribe are powerful diuretics, sometimes expectorants, and in a few instances slightly laxative.

The acids, for in these instances
we are chiefly confined to the vegetable kingdom, are no less distinguished for their taste than for their
properties; but these are sufficient-
ly known. Their opposites, the bitters, scarcely deserve more attention; but there is an obvious difference

ence in the several subjects of this
class. The common warm bitters
are tonic and stimulant; but there
are others which are highly narcotic. Opium is of the latter kind,
and the common bitter, hops, approaches towards it. The tansy,
the wormwood, and many others,
possess this quality in a less degree.
But when the qualities are different, in general the tastes can be
distinguished, though there are
some substances in which these different kinds of bitter seem to meet
so nearly, that it is not easy to say
to which they belong. It has been
supposed, that every substance of
this kind is in some degree narcotic;
for the continued use of bitters fre-

quently

quently deſtroys the tone which they were intended to ſupport or re-ſtore. The purer bitters, the gen-tian and centaury, certainly poſſeſs a ſmall portion of this hurtful qua-lity : thoſe joined with a ſlight a-roma are probably more ſafe and ſa-lutary. The myrrh and the caſca-rilla, the ſimaruba and the colum-ba, both from their taſte and their botanical analogy, as far as has been diſcovered, are of a doubt-ful nature. It is not intended to depreciate their merits, for there is ſome reaſon to ſuſpect that they owe their medicinal virtue to thoſe qualities which we have juſt men-tioned. There is however ſome foundation for the caution which

has

has been given not to continue their use for an extensive period.

The other tastes are the fatty and the styptic. The former is easily distinguished; for it is the unctuous insipidity of pure oil which is generally demulcent, and in a larger quantity slightly laxative. De Haen has endeavoured to extend the virtues of oil, by using it in peripneumonies; but as few stomachs can retain the necessary dose, we cannot enlarge on its expectorant powers. The oil from the liver of the codfish evidently owes its efficacy to the animal matter contained in it, and makes no part of the present subject. The styptic is frequently a compound taste; sometimes, when it

ap-

approaches to the acid, called *au-stere* ; and, when it is nearer to the bitter, *acerb*. They are generally powerful aftringents.

It will be obvious, from this review, that the knowledge which is obtained by the tafte, muft be very general, and capable of fuggefting only a probable opinion of the leading qualities. It would never have pointed out the purgative powers of jalap, or the emetic effect of ipecacuanha. The moft active purgatives, the fcammony, gamboge, elaterium, and colocynth, will either difcover few fenfible qualities, or miflead by thofe which they poffefs. On this account Linnæus had added the *naufeous tafte*, which is the

pecu-

peculiar property of the more violent cathartics, and fome others.
But this depends fo much on different conftitutions, that every fubftance, even in the extenfive catalogue of the ancient Materia Medica, would by many be confidered
as belonging to this clafs. It will
be obvious, that it can be of little
ufe in the inveftigation of the properties of medicines.

There are feven kinds of fmells;
the ambrofial, of which mofch is
an example; the fragrant, or the
flowers of the jafmin; the aromatic,
or thofe of the pink or laurel flowers; the alliaceous, the onion or
affa fœtida; the goatifh, as the orchis, or herb robert; the fœtid, as
the

the hemp and opium ; and the naufeous, as the hellebore and tobacco.
The aromatic and the fragrant feem
to indicate a ftimulating power; but
in general it is flight and tranfitory.
The fœtid and the naufeous are fedative ; but the latter is frequently
emetic and fometimes cathartic.
The goatifh fmell does not point out
any ufeful property ; and the alliaceous agrees in its powers with the
fubftances whofe taftes have been
denominated *acrid.* This is nearly
the fubftance of the boafted knowlege which we poffefs from thofe
fagacious fentinels which are faid to
fuperintend our conduct; and dili-
gently inform us of the approach
of any noxious power. It will not
be

be therefore prefumptuous to fuggeft a method which promifes more
ufeful affiftance; for even the moft.
ftrenuous advocate of the former
fyftems, will either acknowlege
their inefficacy, when generally ex-
tended, or allow that they may with
propriety be ftrengthened by the affiftance of botanical analogy. The
charge of imperfeĉtion may be probably retorted on the prefent attempt—but it is fome merit to have
pointed out a path to general attention, which had been formerly o-
verlooked. This is all our claim,
and we fhall join in the juft applaufe
due to thofe, who by their diligence
will purfue it, and render it more
ufeful.

It

It would be fuperfluous, at this
time, to make any obfervations on
chemical analyfes. The French a-
cademicians have tortured every
medicine to make it confefs its virtues; but with very little fuccefs :
each was obftinately filent, or gave
fuch vague uncertain intelligence,
that the chemift retired in defpair.
We are told by one of this fociety,
that two thoufand experiments had
been tried; and they found only a
little acid, effential, or empyreu-
matic oil, in different proportions;
a fixed and volatile falt; a quantity
of infipid water, and earth. The
very fame proportion of thefe different parts was often found in plants
of very different qualities.

As

As all our guides have been hitherto fo faithlefs, there is little danger in trufting to one, who pretends not to infallibility : in the dark, a blind man may conduct us with fafety; or, if he fails, will at leaft be able to affure us, that it did not arife from the deficiency of his optics. Botanical analogy may indeed fometimes miflead; and in every part of medicine it is of confequence to know, that we do not truft to any thing more certain : but to point out imperfections, is the firft ftep in removing them. In enquiring therefore into the virtues of a plant, it will be neceffary to mention particularly the parts of it which we wifh to examine. Our

com-

common potatoe, though an efculent root, at leaft wholefome, and
once reported fo highly nutricious, and even ftimulating, as to
have excited the following wifh in
Falftaff; " let the fky rain *potatoes,*
" hail kiffing comfits, and fnow *erin-*
" *goes:* let there come a *tempeft* of
" *provocation.*" This common root is
part of a folanum, and the tafte of
the leaves renders it highly probable that they partake of the deleterious qualities of its genus. The
feeds and leaves of the peach-tree
are narcotic bitters; while the fruit
is wholefome and cooling. The
feeds and leaves of the lemon are
bitter; the peel more aromatic; the
fruit acid.

If

If then we examine the place of our more common medicines in their natural orders, we fhall be furprifed at their vicinity, to others of a fimilar quality. The RHUBARB, for inftance, precedes our common dock ; and the old *monks rhubarb* was really taken from the latter ge-nus, — the rumex alpinus. The hellebore ftands in the fame clafs with the aconite, the nigella, and the pulfatilla. The ADONIS, which was probably the hellebore of Hippocrates, joins very nearly to our common hellebore, and to the ane-mone, a plant of fimilar qualities. The folanum, phyfalis, hyofcya-mus, datura, and belladona are in the fame clafs : the digitalis,

lately

lately introduced to the Materia
Medica by the college of Edinburgh, appears to have fimilar qualities. If we enquire into its virtues, from its botanical analogy,
we fhall find that it is probably poifonous, and therefore to be ufed in
fmall qualities;—that it is of the
narcotic kind; but that, as fome
of the clafs are diuretics, particularly the Nicotiana and phyfalis, if it
refembles them, it will probably be
an active and powerful medicine of a
fimilar nature.—Future experience
will determine the propriety and
juftnefs of our fufpicions. The cincona, which produces the Peruvian
bark, is connected with plants,
whofe virtues are yet unknown. The
quali-

qualities of the lonicera and dier-
villa lead us to fufpect that they are
fimilar. They are certainly bitter
and aftringent.

As we have mentioned a few of
the facts to fhew that our purfuit,
like that of many others, is by no
means wholly vifionary, we fhall
confider the different natural orders,
as they occur in the Philofophia Botanica. The firft clafs are called
from their effects piperitæ. They
are all, as far as we are acquainted with them, warm, generally
aromatic, and fometimes corrofive ;
it is therefore probable, that the
faururus and pothos are fimilar. The
latter is commonly called dracon-
tium or arum; the former refem-

bles

bles the ferpentaria. The palm
trees conftantly afford an efculent
fruit, and the medulla is frequently nutricious. From the fcitamina
we have the zedoary, the ginger,
cardamum, amomum, grains of pa-
radife, galanga, and coftus. They
are all warm ftimulauts, though
they differ in degree. The orchis
and its companions are highly nutricious. The iris is the only one
of the following order, whofe virtues we are acquainted with; and
we find its different fpecies æ-
metics, cathartics, or expectorants. But thefe are probably the
confequence of the fame quality, in
different degrees; we fee it in ipecacuanha, in the gratiola, and in an-
timony.

timony. We are not acquainted with the feveral plants with which the fquill is arranged, and confequently cannot compare their virtues; it is certain however that many of them are highly acrid. The roots of the lily tribe are emollient and nutritive. The tulip root is faid to have been eaten ; but not in Holland.. It would certainly be there a capital offence, though the harmlefs Italian indulges himfelf in it, unmolefted. All the graffes are nutricious, except the lollium, of which one fpecies has acquired the epithet temulentum, from its effects; but thefe are loft in dreffing. The leffer feeds are the food of birds ; the greater, of cattle and of men.

The

The following clafs, the CONIFE-
RÆ, are very generally of a fimilar
nature, which will be fully under-
flood, from reflecting that the pine,
the fource of the turpentine, belongs
to it. This very common and ufeful medicine is too much neglected,
at leaft under its proper title; but
the different turpentines, and fome
of the balfams, are very generally
adulterated with the product of the
common fir.

If any one has followed this little
tract with the Philofophia Botanica
in their hands, a degree of diftinction which it will probably never
attain, fo that we need not have
faid a word on the fubject: if however they have done it, they will
find

find feveral claffes omitted; but
they will alfo obferve, that they do
not furnifh a fingle medicine whofe
efficacy is fufficiently decifive to
enable us to compare its companions
with it, and, from the comparifon,
to determine its properties.—Yes,
fays a minor critic, frefh from his
difpenfatory, there is the juncus o-
doratus, a powerful ingredient in
the mithridate and theriaca : but he
ought to be informed that this is
really a fpecies of the andropogon.
It will not be always eafy to anfwer
every caviller, who may think his o-
pinion of confequence ; and the author has little ambition to make a
volume, which would certainly be
the effect of confidering, in a par-

E 4   ticular

ticular manner, *fixty-feven* claffes.
A few of the moft important
circumftances therefore to con-
firm the opinion, which is the object of the prefent tract, will be felected from the numerous facts
fuggefted by the fragments.

The whole clafs diftinguifhed by
the horned antheræ, ftyled BICOR-
NES, are aftringent; and thofe which
bear berries, produce acid and efculent fruit. There feems to be a ftrong
connexion, in vegetables, between
the acid and aftringent principle;
for where the former is obferved in
the fruit, the latter is generally
confpicuous, either in the bark of
the plant, or of the root. It is not
eafily explained; fince, in many inftances,

ftances, a bitter deftroys acid almoft as effectually as an alkali. Some authors have fuppofed, and, if we miftake not, it is an opinion of the Linnæan fchool, that the acid is a component part of the vegetable, obfcured indeed, in the bark, by the fuperabundant aftringency, though again evolved in the fruit. The fact is certain, whatever may be faid of the explanation : the moft ftriking inftances are the vaccinium, erica, and arbutus; but the twenty-fourth order fcarcely contains a plant which properly oppofes the opinion.

Ray, who was rather inclined to think any attempt fimilar to the prefent, a fanciful innovation, yet

allows

allows that his VERTICILLATÆ are generally aromatic ; at least, he observes, that there are more plants of this kind in that order, than of any other. The STELLATÆ of this author, are generally said to be diuretic : the rubia, the aparine, and galium indeed procure this evacuation in a slight degree ; but there is a sedative power, in many of its individuals, which have rendered them useful antispasmodics and anthelmintics. The Indian pink, when recent, is certainly sedative ; and the coffee, though its powers are increased by roasting, is naturally of a similar quality. The famous specific of the Indians against deafness, the *auricularia* of Dale, belongs

to

to this order, and once excited
much attention. Marlow, from
whom Dale received his account,
left no description of the plant, and
Dale only faw it among his fpeci-
mens, and found that its odor re-
fembled that of the water-mint.
Sloane, on this flight foundation,
concluded it to be a fpecies of me-
lifſa. Ray, with little confiftency,
reduced it to the " mentha aqua-
" tica Ceylanica INODORA LATIFO-
" LIA" of Herman, though it was
firft found to belong to the mint,
by its fmell alone. On examining
however the *real* Herbal of Herman,
it was difcovered under the title of
valerianella, with *oblong* leaves ; and
proved to be a fpecies of hedyotis,
and

and to belong to this clafs. This account, taken from Haffelquift, is not a bad fpecimen of the labour of a naturalift, loft in the wilds of conjecture, and a fufficient proof that the moft extenfive knowledge of the fubject, will not fupply the deficiency of real facts. It was this which made Linnæus confider the fimaruba as a fpecies of burfera, the ipecacuanha as an euphorbium, and the jalap as a mirabilis. If he has fo often failed, we may with juftice diftruft ourfelves; yet the fault may be neither his or ours, but the effect of the limited powers of the human mind, which, as well as thofe of the body, require conftant exercife to render them perfect. The naturalift fel-

dom

dom fails in the extent of his me-
mory, or the acutenefs of his ob-
fervation; though he is fometimes
deficient in his reafoning or his
judgment. The hedyotis, fo dear
to thofe who eagerly feek for, and
highly prize fpecifics, is probably
only an antiphlogiftic, flightly
laxative and diuretic, with fome peculiar fedative power. Its ufe in
deafnefs, we can eafily underftand,
fince the chief difeafes of unculti-
vated favages are inflammatory. If
the polifhed European endeavours
to find, in it, a remedy, which will
fecure him from the effects of in-
temperance or diffipation, he will,
as ufual, be difappointed;—but to
return.

The

The asperifoliæ of Ray are slightly aftringent, and frequently mucilaginous. Thefe united qualities have contributed to give them the title of vulnerary ; and, on this account, the lift of Materia Medica is crowded with remedies of a fimilar nature. They are generally of little confequence ; and the judicious phyfician would refign, without a figh, every medicine contained in this order.

The various acrids, or rather ftimulants, which refemble common muftard-feed, alfo refemble each other in their botanical appearances. Even in the fexual fyftem they are connected under the title of Tetradynamia, and in the fragments they are

are called Siliquofæ. The lepidium, cochlearia, raphanus, cardamine, finapi, eryfimum, and fifymbrium, are in general well known. The reft are ftimulants in a lefs degree, and there is no poifonous plant in the whole order.

The Papilionaceæ afford in every inftance, feeds which are the food of different animals; the Columniferi are univerfally mucilaginous; and the Icofandria, which compofe the thirty-fixth, thirty-feventh, thirty-eighth, and thirty-ninth orders, have generally an efculent fruit, without even the fufpicion of poifon, unlefs, perhaps, in the winter cherry.

It

'It is neceffary to mention an obvious, and apparently a powerful argument, in oppofition to the mode of inveftigation, which it is the object of this tract to eftablifh. The cafcarilla—the rival of the Peruvian bark, is found in a poifonous clafs, of which the greater number are draftic purgatives. It would not be difficult to extract from thofe authors, who oppofed its progrefs, when it was introduced by Stahl, as a more fafe and ufeful medicine than the bark, fufficient proofs of its fufpicious tendency. There is however, little probability, that it is highly hurtful. . Its aromatic fumes have produced fainting; but it is very far from a noxious remedy, and

par-

particularly unlike its companions in having no cathartic quality. But though botanical analogy may not be an unerring guide, it may still be ufeful ; and many circumftances may have occafioned an error. It is poffible that its genus may be still miftaken ; or, if it be not, that o-ther parts of its tree may be poifon-ous. The berries of another fpe-cies, the croton tiglium are highly acrid and cathartic ; while its wood, the pavanæ lignum vel Moluccenfe is mild and manageable. Catefby too calls it ricinoides, though the leaf is very different; it is therefore probable that the feeds refemble thofe of the ricinus, which, when

eaten

eaten without any preparation, are highly acrid.

There are many greater difficulties which a fyftem-maker eafily anfwers or eludes; and the fuggeftions juft mentioned may probably be confidered in this light. The fact was mentioned however with a different view, to point out a probable exception; and to fhow, as ufual, that rules can feldom be deemed univerfal. The fufpicious nature of many bitters was formerly mentioned, fo that it is needlefs to repeat them; but there is a natural clafs of plants which are generally bitter, and almoft univerfally aftringent, viz. the dorfiferæ, or ferns, that are juftly fufpected as deleterious.

ous. It may appear ridiculous to argue, that a plant which poifons a worm in the inteftines, may alfo poifon the perfon in whom it dwells; but, independently of the arguments which may be drawn from the general nature of animal life, we fhall probably find all our anthelmintics of a poifonous nature. Arfenic and tin are fufficiently known; mercury and the other minerals will be allowed to poffefs this quality; for poifons do not differ from medicines in their power, but in their dofes. The helleboraftrum or bearsfoot, is a fpecies of Anemone; the fpigelia is fufpected both from its clafs, and its properties, when recent. The antheræ and the feeds

of

of the aquilegia are highly bitter,
and often hurtful. The ferns them-
felves are poifonous to the tænia.
When Madame Nouflers' remedy
was firft publifhed, it was ordered,
in a celebrated hofpital, with the
ufual draftic purgative; but, as
the ferns were not an officinal, and
could not be immediately procured,
the latter was given for fome time
without the former. Many annuli
of a tænia were evacuated, which
were evidently alive;—they moved
fpontaneoufly, and contracted when
ftimulated. But, after the fern-
root was given, they were difcharg-
ed, in a larger quantity, without
any mark of life remaining. The
patient grew cachectic, and was obliged

liged to go into the country. It is not certain that thefe confequences arofe from the ferns; but it is certain, that though a remedy of fome reputation as a fplenic, it is difufed, and exprefly, as is obferved we believe by Vogel, from its having produced fome dangerous fymptoms.

It is not intended that thefe obfervations fhould induce practitioners to defpife bitters, becaufe they may be hurtful. They are certainly generally beneficial in the ufual dofes,' and at proper times. It is however intended to guard them from ufing bitters during an extenfive period, without intermiffion; for, if they poffefs any noxious powers, in this way they will be prin-

F 3

cipally

cipally obferved. If the phyfician purpofes to neglect the narcotic bitters, he will find it very difficult to felect them; for it is highly probable, that fome quality of this kind refides in all. An ingenious friend, who has looked over this little tract, has fuggefted, that, from this quality only, their reputed tonic power may be derived; and that, inftead of reftoring tone, they only deftroy irritability. This pofition muft defend itfelf; and it is only neceffary to add here, that, in the choice of bitters, thofe probably are to be preferred which are combined with an aromatic principle.

It is not only the habit of the plant, the pofition or the ftructure
of

of its flowers, which are fufficient
to fix its place in a natural fyftem :
its fituation and other particulars are
ufually, and with propriety, mentioned ; fo that they ought alfo to
influence the opinion which we
form of its virtues. In moift places,
plants are generally very acrid, and
often poifonous; but the fame fpecies, when removed to drier foils,
are warm and ufeful ftimulants, an-
tifpafmodics or carminatives. It
may feem furprifing, that an ali-
ment fo mild as water fhould nourifh
vegetables, which are peculiarly a-
crid. Yet water is almoft the entire
food of every vegetable ; and it is
moft probable that in thefe, as
well as in animals, what are called

F 4

pro-

production and increase, are only different periods of evolution. The water therefore does not produce 'this acrimony; but peculiar plants, which require a large supply of this liquid, acquire in a moist soil a greater degree of strength and luxuriance, and possess, in their favorits situation, their virtues in greater perfection. We do not want examples of this fact: the arum aquaticum, the colocynth, the Persicaria urens, flammula Jovis, anthora, gratiola, the esula palustris, the ginger, and the eupatorium, as well as a great variety of others which are neglected on account of their too great activity, are chiefly found in moist situations. All the umbel-

umbelliferous plants of this kind,
poffefs very acrid or poifonous powers; but, when removed to a drier
foil, become uieful medicines. The
.galbanum, the opoponax, the
affa fœtida, angelica, cummin, and
mafter-wort are entirely of this clafs.
It will be obvious that the rule,
which is derived from the fituation
of vegetables, cannot be univerfal;
but that highly acrid plants arife
in elevated and dry fituations, while
fome watery ones are mild and mucilaginous. In general, however,
the propofition is true; and, though
there are many exceptions of the
former kind, there are few aquatic
plants which are not acrid in fome
of their parts. There is a more ex-
ception-

able rule derived from authors of real merit, and among the reft from Hoffman, that the ftimulant, aromatic, and aftringent vegetables chiefly flourifh in a dry poor foil; while thofe which are narcotic and poifonous, are chiefly found in rich and moift earth. There are certainly many inftances of both kinds; but there are alfo many exceptions. Yet it is worth recording, as every circumftance which will tend to limit or correct the wanderings of an uncertain traveller, may be fometimes ufeful.

But there is another clafs of plants, which muft ftill be mentioned, though their peculiarity depends neither on their habit or their

I

their fituation, but on their juice: this is the clafs of MILKY plants. They are generally acrid, and fometimes poifonous; the euphorbia, the apocynum, cambogia, and afclepias, are fufficient examples of this quality; but the lettuce, the hawkweed, the dandelion, and fome others, though milky, are much lefs acrid. Thefe are almoft entirely the plants diftinguifhed by Tournefort, under the title of SE-MIFLOSCULOSÆ, and, except a few fpecies, as for inftance the wild lettuce, lately recommended for dropfies, if not mild, they are by no means dangerous. The exception therefore is fo precifely limited,

that

that it fcarcely diminifhes the value of the general rule.

. The facts which have been mentioned, are probably fufficient for the purpofe of the prefent tract, and will induce phyficians to attend more to the botanical analogy than they have hitherto done. It has been obferved, that the attempt to deduce the virtues of plants from their botanical appearances, is as vifionary as to find the philofopher's ftone, or to fquare the circle. It is chiefly to obviate this objection, that thefe obfervations are now publifhed. If the attempt is not entirely groundlefs, more extenfive knowledge, greater opportunities, and more leifure may contribute not

only

only to refcue it from contempt, but to render it really ufeful. From Dr. Hope and Dr. Pultney, much may be expected; and if this little work fharpens their tools, the author's time and attention will not be wholly loft. He will be content to be the mere whetftone, if he can excite fuperior abilities to undertake the work.

In this age of new remedies, the author may probably be indulged with a few reflections on the ftate of the Materia Medica, the accumulated labor of ages, the heap collected from the rubbifh of folly, prejudice, and fuperftition, with a few valuable particles, the refult of real attention and accurate obferva-
tion.

tion. There are two kinds of remedies which form the extremes of the scale; the inactive or inert, and the poisonous. The classes of stimulants and astringents are very extensive; and there is scarcely an object, connected with either, which cannot be amply supplied from those in our hands, and from those in frequent use. It may seem surprising, that many of our new remedies are still of the latter class; we may mention the Quaffia, the Simaruba, the Cortex Indicus Lopezianus, and probably the Columbo. If we except the last, which is a useful medicine in some flow fevers, it is not easy to say what valuable end can be obtained by either of these, thefe, which the common officinals of the fame nature cannot fupply. And even the columbo is often fuccefsfully fuperfeded by the myrrh, or fometimes by the common chamomile flowers. On the contrary, the clafs of poifons have been tortured to afford ufeful medicines, with little falutary effect; yet we now fee the digitalis again recommended, under the fanction of a very refpectable college. We ought not to object to a medicine, whofe properties have not yet been experienced; it has already been mentioned, with fome conjectures relative to its effects, if its very naufeous and acrid tafte will permit its exhibition in a fufficient dofe. The hen-

henbane, the hyofcyamus niger, is indeed an ufeful fedative, and frequently, in large dofes, a laxative; but experience has taught us to difregard the other medicines of this kind, for they are frequently injurious, and feldom ufeful. Even this remedy is frequently defpifed, becaufe its ufual form, the extract, is often injured in the preparation. The feeds are not however liable to this defect; and thofe who will attend to the prefent pamphlet, will attend to the directions of a perfon who has frequently employed them, and generally with fuccefs. It may be neceffary to add, that the firft dofe of the feeds is a grain, which muft be gradually augmented fo far

as

as the head and ftomach will permit ; that the difeafes, to which
they are adapted, are, thofe pro-
ceeding from too great irritability,
particularly in the ftomach and
bowels.

The ufeful medicines are therefore thofe of the middle kind, whofe
effects are neither fo violent as to
endanger the health, which they
are intended to reftore, nor fo flight
as to occafion lofs of time in expectation. If thefe are examined from
their botanical analogy, we fhall
frequently be in little doubt concerning their real virtues. It will
not furprife us that this method has
not yet been found very effectual,
for we lately were ignorant even of

G   the

the true fpecies of the Rhubarb.—
To the younger Linnæus, in his
fupplement, publifhed only in 1781,
we are indebted for the botanical
defcription of the plants which produce the Balfam Peruv.—the Ben-
zoin,—Cubebs,—Jalap,—Ipecacuanha,—Quaffia amara,—Simaruba,
Rhododendron cryfanthemum, —
Santalum rubrum, — the common
Nutmeg, and fome others. To the
late Dr. Fothergill,—to Mr. Ker of
Patna, and to other attentive obfer-
vers in India, Carolina, and Jamaica, we owe much information within thefe fifteen years; fo that this
fubject is ftill almoft in its infancy;
and the natural method of arranging
plants, has been more admired than
culti-

cultivated. We are yet ignorant of many plants which have afforded us useful medicines, as the Ammoniacum,—Myrrha,—Gum Kino, &c. and there is much reason to suspect, that those already supposed to be known, have, in some instances, been mistaken, and really belong to very different genera.

The number of remedies have been considerably increased by the eagerness of physicians to find specifics. To the untutored savage we owe several valuable ones; and we fondly cherish the delusion, that he has still greater treasures in store. It is necessary therefore to observe, that no remedy deserves that title; and, when we are better acquainted

G 2   with

with thofe which have attained it,
we generally find · that they act on
general principles, and that their
virtues differ only in degree from
thofe which we have conftantly employed. In the Amænitates Acade-
micæ, we have however a differta-
tion on the fpecifics of the Cana-
dians, which contains near forty
fpecies. We ought not to fay that
thefe are not worth our attention ;
there may be fome valuable remedies among them, and they deferve
farther examination ;—but we can
venture to prophecy, that every in-
dividual will be readily reduced to
the claffes we are already acquainted with, and that no one real fpe-
cific will be found in the number,

Reme-

Remedies of this peculiar nature, are the very foundation of every empirical fyftem, and they are the refuge of ignorance and idlenefs. It is probable therefore, that our charms will not have much effect. Specifics will ftill be the ignis fatuus which will miflead the accurate enquirer, and perplex the rational phyfician.

But the materia medica, as well as medicine in general, has fuffered from different and oppofite errors. If the virtues of remedies have been fometimes too much confined, at others, they have been confidered as too general; and phyficians have formed claffes, before they have examined individuals. The arrange-

ments

ments of the earlier authors, were in
many refpects defective ; thofe of the
lateft and moft judicious moderns are
fo comprehenfive, as frequently to
miflead, or, at leaft, to afford little
fatisfaction.   If a medicine is claff-
ed from its effects in health, it muft
often be arranged very differently
from its effects in difeafe ; and probably its title will fometimes vary,
as the difeafe changes. Blifters, for
inftance, ftimulate, yet they frequently produce fleep ; and opium
is a fedative, though it often in-
creafes pain.   There is therefore no
criterion, by which we can always
determine the place of a remedy ;
and the ftudent who only confults
his fyftem, will frequently produce
effects

effects very different from thofe
which he wifhed for. Another objection will alfo occur to an attentive obferver; viz, that medicines
often poffefs more than one quality;
and, though the principal and pre-
dominating effect generally charac-
terifes the remedy, yet, in many
inftances, it is not eafy to difcover
it. Camphor we know, is frequently an ufeful expectorant; in other
fituations a proper fedative; yet in
the end of fevers it appears to be
a neceffary cordial. In what clafs
then, can it be with propriety arranged? Among the fedatives? It
will not eafe pain, it will not procure fleep, it will not always check
any extraordinary difcharges. A-

G 4   mong

mong the expectorants ? We shall
in vain trust it in the fit of an asth-
ma, in a serous catarrhal defluc-
tion, or in the confuming ulcer of
a phthisis. It is ufelefs to multiply
inftances; for, though this be a
very peculiar medicine, yet almoft
every other, when examined with
precifion, will afford equal difficul-·
ty. The fum then of the whole is,
to recommend an attentive examination of individuals, and to begin
with lower orders, rather than at
once to form extenfive claffes; to
arrange a few together, whofe properties are fimilar, and to connect
the higher arrangements by their
more general agreements; in fhort,
to form genera and orders, before we
afcend

afcend to claffes. We have already given fome inftances of this kind in our obfervations on aftringents. The columba root and myrrh feem to agree in diminifhing irritability; the cafcarilla and bark, in reftoring tone; and both are adapted to fevers in their different forms. Again, there are fome aftringents which are ufeful vulneraries; and there are others, nearly allied to them, which are powerful ftyptics; not to add, becaufe it has been already infifted on, that the limits which divide the aftringents from fedatives, are narrow and inconfiderable. In the lift of fedatives, there is a great difference between thofe gums which correct

flatu-

flatulence and convulfive motions in the bowels, and opium, which, in large dofes, feems to check all the vital and animal actions; between the caftor, which fometimes relieves the fpirits of an hyfterical female, and the copper which frequently checks the violent diftortions of the epileptic.

The great variety in the properties of medicines, will not be eafily explained ; and a confiderable extent is particularly improper in this flight view, which is rather calculated to rouze and direct phyficians, than to conduct them. If thefe obfervations are well founded, they will ftrongly fupport the alphabetical, botanical, or any arbitrary order

der of defcribing the virtues of medicines. If any other be allowed,
it muft be that adopted by Dr. Cul-
len, whofe introductory obfervations to each clafs, and the arrangement of individuals under it, are
frequently calculated to remove the
objections now mentioned. Indeed
there are fo many pofitive advantages in his method, though the execution is imperfect, that we only
wifh to guard it from error, and to
conduct it with fewer difadvanta-
ges. It is for a time only that we
fhould return to arbitrary methods,
and ftudy Linnæus, Lewis, or Ber-
gius: it is only to be more perfect
in our elements, before we attempt
intricate combinations. We ought
not

not to conceal, that fome efforts of this kind have been made by a very candid and intelligent author; we mean Dr. Duncan, whofe work would have been more perfect, if his materials had been more fatisfactory. If his orders are fometimes lefs ufeful and interefting, it muft be attributed rather to the ftate of the fcience, than the deficiencies or errors of the author.

In this purfuit, we ought not to neglect combinations of medicines. From the more ancient phyficians we receive little fatisfaction, becaufe, in the mafs of remedies, we can feldom feparate the efficacious from the inert, the ufeful from the trifling. It is probable that, as u-
fual,

fual, the moderns have fell into the oppofite error, and rendered that fimple, which would fometimes be more ufeful, if more complicated. It is not intended to plead, in the moft diftant manner, for the con-fufed mafs which once filled the prefcriptions of phyficians. On the contrary, the real effects of reme-dies can only be known, by ufing them fingly and feparately; but it is necelfary to fuggeft, that the practitioner fhould not confider this, as the acmè of his art. If com-pounds have really different virtues, they fhould be inveftigated, and properly employed; fo that even the apparently heterogeneous mix-ture of folly or fancy, if its virtues

have

have not been tried, fhould not be contemptuoufly neglected. Dovar's draftic purge, his fudorific powder, Fuller's balfamic pills, and fome other ufeful remedies are ftriking proofs of the different power of compounds.

It is alfo probable, that external applications have not been fufficiently attended to. In the ileus, every medicine is rejected. In fome irritable habits, emetics produce violent diforders ; and, when poifons have been taken into the ftomach, we are fearful of increafing the irritation, though oil and warm water have little effect in difcharging its contents. If it be true, that hellebore applied to iffues has produced
duced

duced an evacuation of the intefti-
nal canal, it is probable, that when
ufed in a fomentation to the bowels,
it would be more certainly effectual.
If it be true, that common groun-
fel, applied to the ftomach, will
produce vomiting, it may be frequently ufeful, to thofe who can,
with difficulty, bear any medicine
of that kind ; or, as in the former
inftance, where we fear to add another irritating fubftance. There
are other inftances of this fort ; but
this is not the proper place to enlarge on them.

Thefe few obfervations may not
be thought impertinent, in a fubject firft fuggefted by an attentive
confideration of the Materia Medica.

ca. At some future period, they
may be enlarged; so as to be more
worthy of attention. In this curious age, it will probably be of service to check, rather than to excite; to step forward to the eager
projector, and with a calm insinu-
ating voice to restrain his ardor.
" Do not be too impatient, in your
attempts to improve : examine the
resources in your hands : enquire
into their defects, and learn whether your boasted novelty can claim
greater virtues, or be disgraced by
fewer faults. If it be the same, let
it not be encouraged, only because
it is procured with difficulty and
danger, or because it is now first
introduced. If it be worse, reflect
that

that your feducing promifes may
endanger the life of a valuable citi-
zen, or an ufeful member of fociety.
Above all, reflect that the reputation which you feize by violence,
is the tranfitory gleam which will
foon vanifh, and leave you in your
former obfcurity;—that the only
reputation really defirable, is that
folid, well-grounded confidence,
which, as it is not the effect of a
momentary bubble, will be pro-
portionably more durable : a reputation not founded on the credit of
a fallacious remedy, but the confequence of thofe offices of fkill and
humanity, which add a dignity to
the phyfician and the man."

H   Remon-

Remonftrances and reafon will, perhaps, have, in this inftance, an equal effect, and be liftened to with indifference, or foon forgotten. New remedies will ftill be offered; we fhall ftill expect miracles, and be repeatedly deceived. It might add fome efficacy to this tract, if the Materia Medica were examined in all its varieties with care and attention, with a coolnefs of invefti-gation, and accuracy of difcern-ment; fo that while every fource of information is impartially at-tended to, even the moft refpectable is rigidly fcrutinized. Truth fhould be diftinguifhed from doubt, and probability from miftake, as far as experience has already led us.

In

In this inveftigation, we fhould clearly fee our own negligence and errors, while the refult would be, a cautious referve, or a rational fcepticifm, when new remedies are propofed. The vaft extent of the fubject would be no rational objection, except to the labor of the undertaking, for facts are eafily diftinguifhed from opinions, and mankind only difpute about the latter. Yet even opinions fhould be regarded, fince in medicine, there are few undifputed truths; and the controverfies of pedants, and empirics, as well as thofe of attentive and rational phyficians, fhould not be neglected. But the refult only may be recorded, with proper reference to

the

the original authors. In this way
a valuable mafs may be accumulated, not indeed, with little labor,
but with much information; not
enormous in bulk, but rich in matter. If this work be attempted by
able hands, we fhall receive great
fatisfaction from it; if it be not, its
utility may be difcerned by an inferior execution.

It is now neceffary to conclude,
left what was intended for a fketch,
fhould become a volume, and the
running wheel produce an hog-
fhead, when we promifed only a
pipkin. This fubject may probably, at fome future period, be ex-
tended; and what was propofed only
to be a floating fheet, affume a more

refpect-

refpectable form. "But we are
afhamed to be inftructed by thofe
with whom we are not acquainted."
True : yet the author and the reader
muft, for a time, continue to be
ftrangers to each other. If a more
intimate acquaintance promifes
greater entertainment, the mafk
may drop, and what was begun in
obfcurity, be continued in open day.

F I N I S.

www.ingramcontent.com/pod-product-compliance
Lightning Source LLC
Chambersburg PA
CBHW021942190326
41519CB00009B/1114